Kleuton Antunes Lopes Lima Antunes
Heurison Silva

Molecular Modeling of Tartrazine in Aqueous Media

Kleuton Antunes Lopes Lima Antunes
Heurison Silva

Molecular Modeling of Tartrazine in Aqueous Media

Comparative study between the UFF and Ghemical force fields in the azo(-N=N-) group in the modeling of Tartrazine

ScienciaScripts

Imprint
Any brand names and product names mentioned in this book are subject to trademark, brand or patent protection and are trademarks or registered trademarks of their respective holders. The use of brand names, product names, common names, trade names, product descriptions etc. even without a particular marking in this work is in no way to be construed to mean that such names may be regarded as unrestricted in respect of trademark and brand protection legislation and could thus be used by anyone.

Cover image: www.ingimage.com

This book is a translation from the original published under ISBN 978-3-330-77104-8.

Publisher:
Sciencia Scripts
is a trademark of
Dodo Books Indian Ocean Ltd. and OmniScriptum S.R.L publishing group

120 High Road, East Finchley, London, N2 9ED, United Kingdom
Str. Armeneasca 28/1, office 1, Chisinau MD-2012, Republic of Moldova, Europe
Managing Directors: Ieva Konstantinova, Victoria Ursu
info@omniscriptum.com

Printed at: see last page
ISBN: 978-620-8-63540-4

Summary

ACKNOWLEDGMENTS

To God, what would I be without the faith I have in him?

"Give the world the best of yourself. But that may not be enough. Give the best of yourself anyway. You see, at the end of the day, it's all between YOU and GOD. It was never between you and others."

(Mother Teresa)

Summary

In this work, molecular modeling of tartrazine was carried out in order to analyze the molecular conformation with emphasis on the azo group, considering the energetic and geometric parameters. To do this, the molecule was subjected to the UFF and Ghemical force fields using the Avograd program. Both in isolation and in an aqueous medium (surrounded by water molecules). The results obtained for the UFF field in the analysis of molecular mass and energy have the same behavior as for the Ghemical force field. However, the distance between the azo group behaved differently: the UFF force field describes a drop of up to 20 water molecules and for larger values this distance reaches an average value of 3.6545 angstrom. For the Ghemical field, this drop occurred up to 30 water molecules. From this value, it oscillates around 3.544 angstrom. Comparing the distance between the carbons containing the azo group for the two force fields, the Ghemical field decreases the distance by around 2.8%, proving that the interactions in this field affect geometric conformity by reducing the azo coupling and , thus affecting the instability of the molecule and potentially changing its color. For the angles α and β, the UFF field shows us an expected symmetry, as this field disregards intermolecular interactions, which is not the case with the Ghemical field.

Key words: Modeling, Molecule, Tartrazine.

Chapter 1: Literature review

1.1. Dyes

A dye is a chemical compound that can be attached to a material, giving it a certain color, due to the presence of chromophore groups in its structure. Chromophore groups are characterized by having conjugated double bond systems, the most common of which are:

$>C=C<$, $>C=N-$, $>C=O$ and $-N=N-$.

The first dyes were used by humans in prehistoric times, in what are now known as cave paintings. These were natural dyes, obtained from a natural extract of a plant or animal. With the passage of time and the increasing use of these dyes, Willian Henry Perkin synthesized a dye for the first time, which triggered a breakthrough in several sectors.[1]

Today, these dyes are found on a large scale in the textile, food and pharmaceutical industries. However, some people are more sensitive to the use of these substances, developing a series of problems such as allergies, asthma, urticaria and others. This has led to a series of studies such as these, which aim to find out more about dyes.

1.2. Why study dyes

Colors are closely linked to various aspects of people's lives and are capable of influencing everyday decisions, especially those involving food.

Color is one of the first sensory attributes that influence the acceptability of a given product, especially with the constant growth of the industry, which uses this mechanism as a way of attracting customer attention and increasing its resources. In this way, the judgment of this parameter is important in terms of its components and its physical and chemical properties, so that it can even help many people who are constantly allergic to these products.

1.3. Advantages of synthetic dyes over natural ones

Synthetic dyes have advantages over natural dyes in that they provide a wide variety of colors, are more stable to certain factors such as pH, light and temperature, have a high absorption power and a low production cost. For this reason, they are used more easily, the industry would have lower production costs, due to the wide variety of synthetic colors compared to natural colors, greater attractiveness to some eyes, and the greater difficulty in obtaining and maintaining natural dyes.

1.4. Tartrazine

Tartrazine is a chemical compound that gives food its yellow color. It is widely used in industry because, in addition to giving food its color, it has the function of attracting consumers to the visual aspect of a product that contains this chemical compound in its composition.

In some countries, the use of this chemical compound has been restricted because some studies have shown that it has caused human impairment. In Brazil, its use is restricted by Anvisa.

Figure 1: Molecular structure of tartrazine.

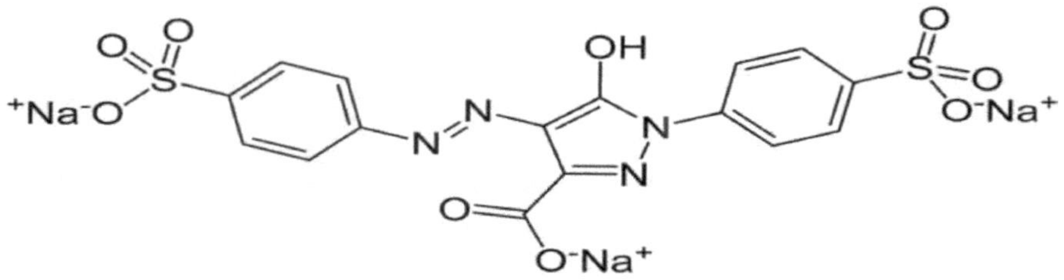

1.5. Technological implications related to the use of colorants

One of the major concerns related to the use of colorants in foods is maintaining their chemical stability. Due to the complex diversification of food matrices, the chemical environment in which the dye will be exposed is one of the main factors impacting on the maintenance of the chromophore's molecular structure and, consequently, on the regularity of the color in the food.

In this work, we will learn how to use the Avogadro program (http://avogadro.openmolecules.net/wiki/Main_Page) to model the tartrazine molecule. We will consider the various force fields allowed by the program, in particular UFF and Ghemical[1] to check the molecular conformation when

1 For more details on force fields, see section 2.2. Some force fields used in modelingcomputer

subjected to different situations. The work focuses on an analysis of the behavior of the chromophore group when the molecule is in its normal state and inserted into aqueous medium. By varying the number of water molecules, we are changing the physical and chemical environment of the molecule and, consequently, modifying the shades. Initially, we kept the molecule in a non-interacting state (isolated molecule) and then changed this state to one of molecular interaction in an aqueous medium.

Chapter 2: Molecular modeling briefly presents the main characteristics of computational modeling and the important parameters for applying the technique, as well as some characteristics of the interaction algorithm that supports the optimization of the program's molecular geometry.

Next, in Chapter 3: Methodology, we describe the methodology used to carry out this work, step by step (Section 3.2). We also describe the program (*software*) used here, Avogadro®, which can be obtained free of charge (in demo version) at http://avogadro.openmolecules.net/wiki/Main_Page. It is worth considering that even in the free version, satisfactory results can be extracted with regard to the objectives proposed here.

The results and discussions are presented in Chapter 4: Results and data analysis, and the conclusions and prospects for future work are shown in Chapter 5: Conclusion.

Chapter 2: Molecular modeling

This chapter presents molecular modeling techniques, their main advantages and some applications.

2.1. Molecular modeling

Molecular Modeling (MM) comprises a number of computational and theoretical tools and methods that aim to understand and predict the behavior of real systems; used to describe and predict molecular structures, transition state properties and reaction equilibria, thermodynamic properties, among others.

These methods include energy minimization studies of molecules, conformational analysis and molecular dynamics simulations; they are applicable to isolated atoms and biomacromolecules.

Molecular modeling basically involves four stages:

(i) Select a model that accurately describes the inter- and intramolecular interactions of a system;

(ii) Perform the calculations;

(iii) Analyze the results, validating or rejecting the chosen model.

which are supported by three variables (α, β and C) being analyzed a priori:

(iv) The size of the system to be studied (in terms of number of atoms);

The accuracy desired in the results (varies according to the method chosen to calculate a given property), the computational cost and the hardware conditions used to carry out these calculations. In addition to providing structural data, theoretical calculations are also used for chemical and pharmacological purposes, such as computing heats of formation of molecules, interatomic distances, electronic energies of HOMO and LUMO, ionization energies, atomic electronic densities, net atomic charges, bond orders, dipole moments, among others. The great development of MM is largely due to the advance of computing resources in terms of hardware and software. In the past, the use of MM was restricted to a select group of people who developed their own programs. Nowadays, it is no longer necessary for a "modeler" to compose their own program because they are marketed through large companies and even academic laboratories. Molecular modeling provides important information for the drug planning process. It allows specific properties of a compound to be obtained which can influence the interaction with its receptor. Examples include the electrostatic potential map, the electronic density contour and the energies and coefficients of the HOMO and LUMO frontier orbitals. Other important information can also be obtained from the structural comparison between different molecules, which allows the generation of a similarity index that can be correlated with pharmacological activity.

In this work, we used the Avogadro® program, which can be downloaded from the international computer network and has the advantages of easy handling and the possibility of using different configurations or force fields. Here, we have chosen to use the *Universal* Force Field (UFF), as this is the most promising force field today.[2] In addition, we will also apply the Ghemical force field in order to compare the geometric quantities of the molecule in isolation and also in interaction with the medium.

The difference between the UFF force field and the Ghemical force field is due to the interactions between the molecules. In the UFF force field, these molecular interactions are negligible, while in the Ghemical force field they exist according to the interaction algorithm that will be mentioned in section .

2.2. Some force fields used in computer modeling

There are several force fields available for the most varied applications. The Avogadro® *website* itself provides tutorials that make it easier to understand the program, manipulate it and choose the appropriate force field. For example, chemists are more interested in visualizing the molecule in terms of the length of the bonds and the angle between them, which naturally depends on the coordinate system adopted.

Spectroscopists, on the other hand, are more interested in determining the geometric parameters and force constants of the bonds that give results that coincide with the experimental data. It is also important to consider that each force field has its own limitations.[3]

Although there are many force fields in use, such as Dreiding, MM1, MM2, AMBER (acronym for *assisted model building and energy refinement*), OPLS (*Optimized Potentials for Liquid Simulations*), Johnson's force field, etc. the following is a brief presentation of the force fields available in the *software* used by us, their main characteristics and applications. More information can be found on the *internet* as well as in specific bibliographies.4 However, it is not our intention to do a detailed review of all the literature concerning the available force fields, but only to illustrate their main applications and the most widely used to date.

2.2.1. Ghemical

The Ghemical force field is most commonly used to model simple organic molecules . This resource constitutes a class of force fields used by the Ghemical® program.[13] More information can be found at http://openbabel.org/wiki/OBForceFieldGhemical.(5) The parameters analyzed for this field are:

a) Binding energy:

Figure 2: Graph of the binding energy between atoms (from reference 5)

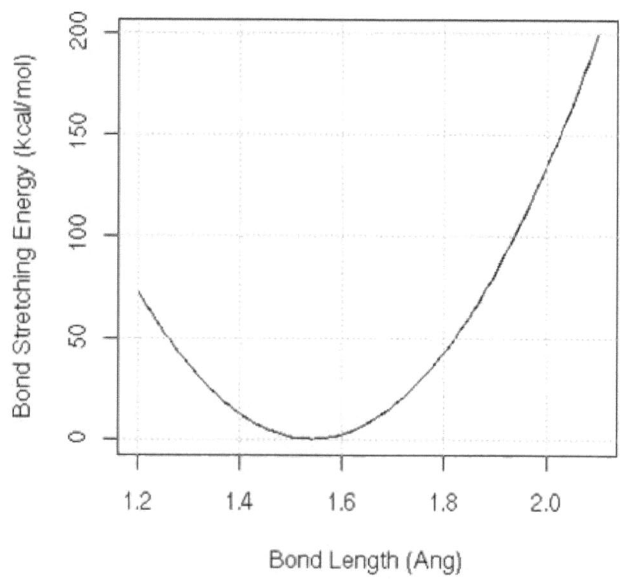

$ab-¿r^{0}_{ab}$

$r_{¿}b$, K_b

where $E_{Bond}=K_b¿$ *is the* bond length of the force constant, r_{ab} is the bond distance between atoms A and B, and r^{0}_{ab} is the bond equilibrium point between atoms A and B.

b) Angular bending energy:

Figure 3: Graph of angular bending energy between atoms (from reference 5)

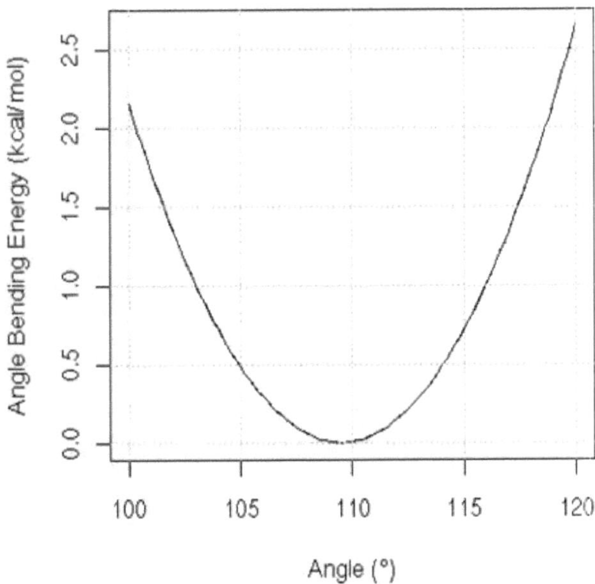

where $E_{angle}=K_a(\theta_{abc}-\theta_{abc}^0)^2$, K_a is the force constant, θ_{abc} is the angle, and θ_{abc}^0 is the ideal angle.

c) Torsional energy:

Figura 4: Graph of torsional energy and its local derivatives (from reference 5)

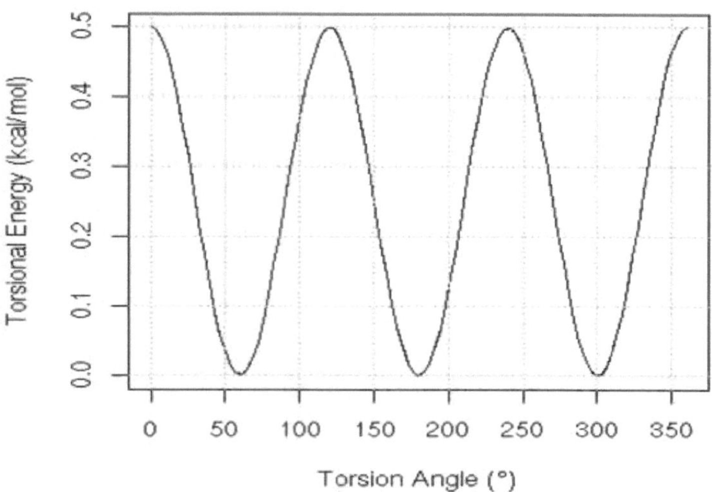

	$s_t n_t = +3$	$s_t n_t = +2$	$s_t n_t = +1$	$s_t n_t = -1$	$s_t n_t = -2$	$s_t n_t = -3$
V_1	0	0	V_t	$-V_t$	0	0
V_2	0	$-V_t$	0	0	V_t	0
V_3	V_t	0	0	0	0	$-V_t$

Where $E_{torsion} = V_1(1+\cos[\omega_{abcd}]) + V_2(1+\cos[2\omega_{abcd}]) + V_3(1+\cos[3\omega_{abcd}])$, V_t is the rotational barrier, S_t is +1 for a local maximum or -1 for a local minimum. n_t is the multiplicity, $\omega_{(abcd)}$ is the torsion angle and V_1, V_2 *and* V_3 = Derived from $V_{(t)}$, $S_{(t)}$, $n_{(t)}$used in the table Figure 4 .

d) Van der Waals energy:

Figure 5: Van der Waals energy graph (from reference 5).

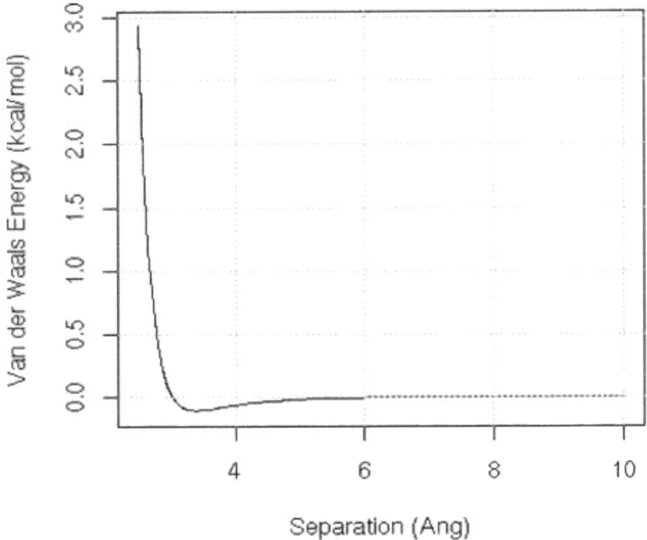

Here, $E_{vdw}=K_{ab}\left(\frac{1}{\sigma_{ab}^{12}}-\frac{2}{\sigma_{ab}^{6}}\right)$, $K_{ab}=\sqrt{K_a}\sqrt{K_b}$, $\sigma_{ab}=(\frac{r_{ab}}{r_a+r_b})$, r_a *is the* atomic radius *and* r_b *is* the separation between the atoms.

e) Electrostatic energy:

Figure 6: Electrostatic energy graph for (q=+0.1,q=-0.1) (from reference 5)

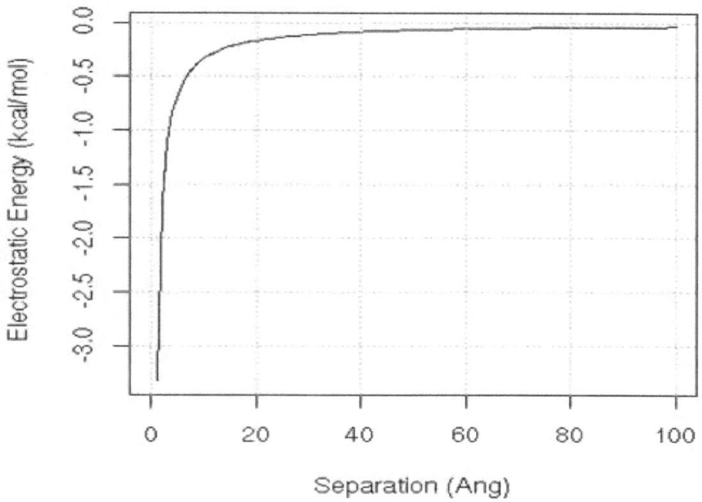

where $E_{ele} = 332.17 \frac{Q_a + Q_b}{r_{ab}}$, Q_a is the atomic charge (atom A), Q_b is the atomic charge (atom B) and r_{ab} is the separation between atoms A and B.

All of these equations are relevant to the Ghemical force field.5

2.2.2. MMFF94(s)

It is the force field used to model drug molecules and is also widely used in organic chemistry. This force field is most appropriate for alkanes, alkenes, alcohols, phenols, ethers, aldehydes, ketones, acetates, ketates, hemicetates, hemiacetates, amines, amides, peptide analogues, ureas, imides, carboxylic acids, esters, anions, carboxylates, ammonium cations, thiols, mercaptans, disulfides, halides (chlorides and fluorides), imines, iminium cations, amine N-oxides, hydroxylamines, hydroxamic acids, amidines, guanidines, amidinium cations, guanidinium cations, imidazolium cations, aromatic hydrocarbons, and heteroaromatic compounds.[6].

Both the MMFF94 and its variation, the MMFF94s, use the same functional form to calculate potential energy. They only differ in the parameters of *torsion* and *out-of-plane* bending. The 's' in Mmff94s stands for *static* and this set of parameters is best suited to tasks where the result is static. To understand what this means, just remember that a result such as an energy minimization is static (only the geometry is obtained). On the other hand, a dynamic result would be the trajectory of a molecular dynamics simulation.

2.2.3. UFF

This force field can be used throughout the entire Periodic Table. It is described as being made up of parameters estimated using basic rules for each element in the Periodic Table, such as its hybridization and connectivity. In 1992, Rappé *et al.* presented a paper describing the functional form, parameters and generating formulas of this field for the complete Periodic Table.

In general, the parameters used here include an adjustment of the hybridization-dependent atomic radii, an adjustment of the hybridization angles, parameters concerning the van der Waals interaction, torsion conformation, and the effective charges of the atoms. [(21)]

This was the force field used in this work, as it disregards other interactions that are not directly associated with the atoms of the molecules we are interested in.

2.3 Interaction algorithm[2]

The interaction algorithm is the "heart" of the molecular modeling program. It is the mechanism responsible for determining the path by which the optimized structure and minimum energy are obtained.

In molecular mechanics, energy equations are used to calculate the interactions and the corresponding energies of the mechanism, as it is impossible to simulate reactions in molecular mechanics. In other words, bonds cannot be formed or broken.[7,8]

A force field works as a set of equations and parameters that define these equations. Every simple force field must express the energy as something like the equation below:

$$E_{total} = E_{ligação} + E_{ângulo} + E_{Torção} + E_{vdw} + E_{Eletrostático}$$

2 According to the text of Reference 11.

Following the laws of classical physics. Here, atoms are represented as particles (nuclei), without considering electrons. Atoms are joined together by bonds.

Each of these individual energy terms in Equation 1 above needs parameters that depend on the types of atoms for the interaction. The term refers to the energy associated with van der Waals interactions. A more detailed description can be found in reference[9] Below, we present some of the algorithms used in molecular modeling, available in the Avogadro® program.[10]

2.3.1. *Steepest descent*

Steepest descent is an iterative optimization algorithm. It works repeatedly, moving the atoms to a local minimum in the direction of the forces. This means that the instructions for successive steps are orthogonal. As a result, the steepest descent algorithm converges rather slowly in narrow valleys on the energy surface. *Steepest descent* is well suited to an initial minimization to relieve the molecule from large forces. This algorithm was used to optimize the geometry of the molecules studied[11].

2.3.2. *Conjugate gradients*

Conjugate gradients is another iterative optimization algorithm. It works by repeatedly moving the atoms to a local energy minimum in the d_i direction:

$d_{i=} -g_i + \Upsilon_i d_{i-1}$

com $\Upsilon_i=(g_{i}. g_{i}/d_{i-1}. d_{i-1})$.

Here g_i is the gradient for the actual step, g_{i-1} is the gradient for the previous step. Since it doesn't exist at the start of the process, the first step is identical to the first step for the Steep Descent. The Conjugate Gradient theoretically converges in the same number of steps as the variables (3N, where N is the number of atoms). However, the line shape never finds a true minimum. For this minimum to be effectively reached, some additional steps are often necessary.[12]

Chapter 3: Methodology

3.1. The program

The Avogadro® program is easy to use and is available free of charge at http://avogadro.openmolecules.net/wiki/Main_Page. Figure 7 shows the Avogadro® program window during the modeling of the adenosine triphosphate (ATP) molecule. The energy value is shown during the molecular structure optimization process in addition to the value of the dE derivative. The process stops when dE equals zero, indicating the

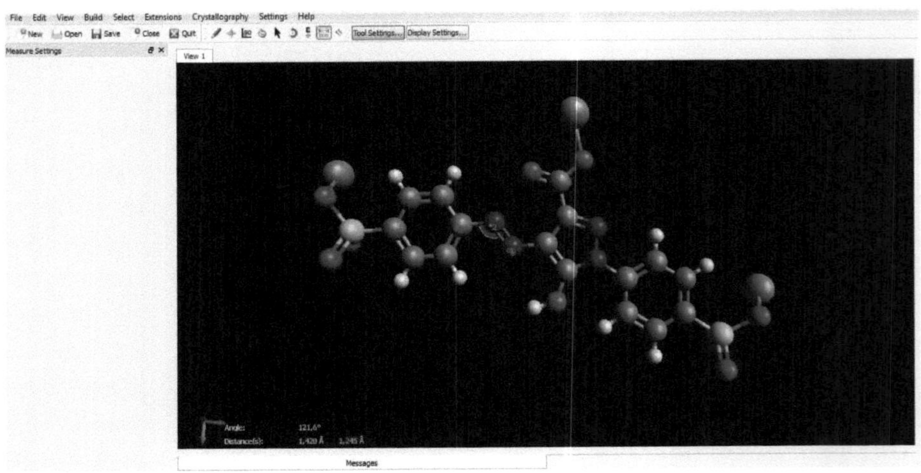

minimum energy value.

Figure 7: Avogadro® program screen during the optimization of the tartrazine molecule

The program allows you to interfere with the system at any time, changing atoms (one or more) or even the configuration (relative position) of one or more constituents of the system. It is also possible to add electrical potential gradients such as flat or cylindrical charged surfaces. The Avogadro® program has an easy-to-use interface, which allows you to choose the appropriate force field according to the purpose or system under study in an intuitive way, the software allows you to create molecules using your *mouse*. Simply click on one position and drag to a second to obtain a bond and two molecules. The user can change the atom to be used and the quality of the bond between them. You can also make various changes such as names, annotations, angles and so on.

A highlight of this program is that it allows the molecules created to be viewed in three dimensions, and for those who prefer some dynamism, you can also choose to view them rotating in an animation. Visualization options and other ways of filling in the lines can be applied in the options on the right-hand side of the program screen. Read more in: (see Section 2.2. Some force fields used in computer modeling).[13]

3.2. Procedure

In the molecular modeling of tartrazine, we tried to analyze the behavior of the azo group when water molecules are added to its environment. The parameters analyzed in the modeled molecule are: C, α and β, where C is the distance between the carbons in which the nitrogens are double-bonded, α and β are the angle between the double bond made by the nitrogens and the bond between the carbon. See Figure 8 below:

Figure 8: Tartrazine molecule with the parameters analyzed in this work

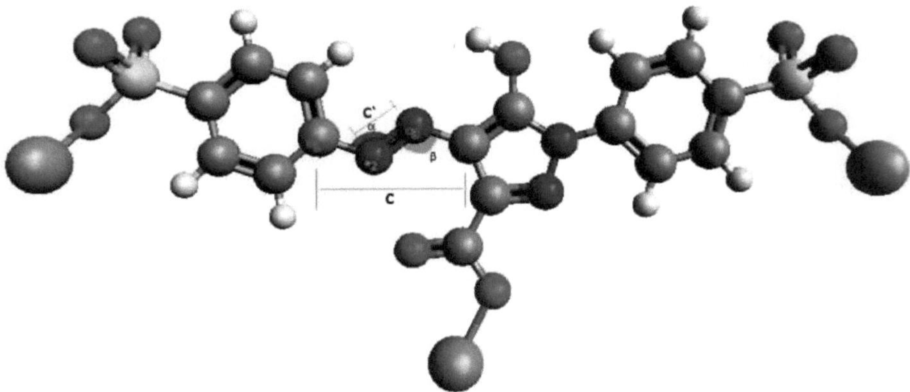

Using the Avogadro program, we modeled the tartrazine molecule and simulated it in an aqueous medium. This was done for two different force fields: first we used the UFF force field, which has already been described in section 2.2.3. UFF, we analyzed the parameters shown in Figure 8 and compared them due to the addition of water molecules, as we increased the number of water molecules we noted the distances C between the carbons containing the azo group in their environment. We also noted down the angles α and β, and the distance C', which is the distance between the nitrogens of the azo group, in order to check for possible changes in the molecular structure. Figure 9 below shows the tartrazine molecule with 40 molecules of water to illustrate how the modeling was carried out to obtain the data. The same procedure was carried out with the application of the Ghemical force field.[14]

Figure 9: Tartrazine molecule in an aqueous medium.

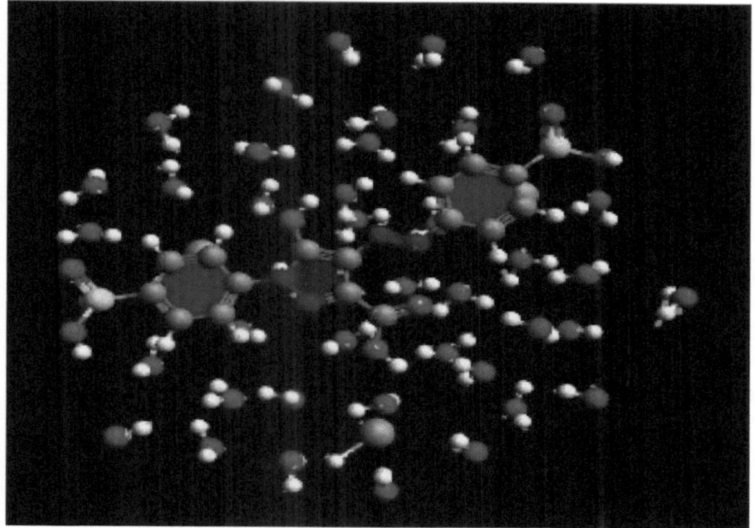

Chapter 4: Results and data analysis

4.1. The tartrazine molecule

The results obtained from the modeling of tartrazine for Error: Reference source not found are shown in Table 1 and Table 2. In them, we can see the main information that must be taken into account regarding the structure of the molecule, such as the minimum energy of the configuration, the angle of the bonds between the chromophore group (- N = N -) and the binding carbon, which I have called α and β, as in Figure 4 the molecular mass, the angstrom distance of the double bond of the azo group, and the distance between the carbons that comprise the azo group. With this data, we will be able to observe the behavior of the azo group if we change the chemical environment in which the tartrazine molecule is found.

However, one quantity that is of great importance is the molecule's optimization energy, which is associated with the best molecular conformation. For isolated tartrazine, we obtained a value of 1198.583 kJ/mol in our simulations in the UFF force field. This is the energy value in kJ/mol of the molecule in its natural state, without imposing a chemical environment that modifies the molecular structure through chemical interactions.

4.2. Analysis of the azo group at UFF

Table 1 below shows the results obtained by optimizing the tartrazine molecule in the UFF (*Universal Force* Field), with the addition of water molecules in its vicinity. This data has been conveniently displayed in the form of graphs, as shown below.

Table 1: Tartrazine molecule data (force field - UFF).

Tartrazine	Energy (Kj/Mol)	Angle		Molecular mass (G/Mol)	Distance (-N=N-) r	**Distance between** c_1 **and** c_2
		α	B			
Tartrazine	1198,58 3	121,6°	122,4°	534,363	1,245	3,657
With 10 molecules of water	1109,24	121,6°	122,4°	714,516	1,244	3,654
With 20 molecules of water	1043,35 1	121,5°	122,4°	894,669	1,244	3,653
With 30 molecules of water	965,817	121,5°	122,5°	1074,822	1,244	3,655
With 40 molecules of water	942,852	121,7°	122,3°	1254,975	1,244	3,654
With 50 molecules of water	885,68	121,7°	122,3°	1435,127	1,244	3,654
With 60	801,576	121,7°	122,2°	1615,280	1,244	3,654

water molecules						
With 70 water molecules	740,035	121,5°	122,5°	1795,433	1,244	3,654
With 80 molecules of water	655,21	121,7°	122,3°	1975,586	1,244	3,655
With 90 molecules of water	593,522	121,7°	122,3°	2155,739	1,244	3,655
With 100 molecules of water	569,076	121,4°	122,6°	2277,81 4	1,244	3,655

Graph 1 : Energy and molecular mass of the system as a function of the number of water molecules surrounding the tartrazine.

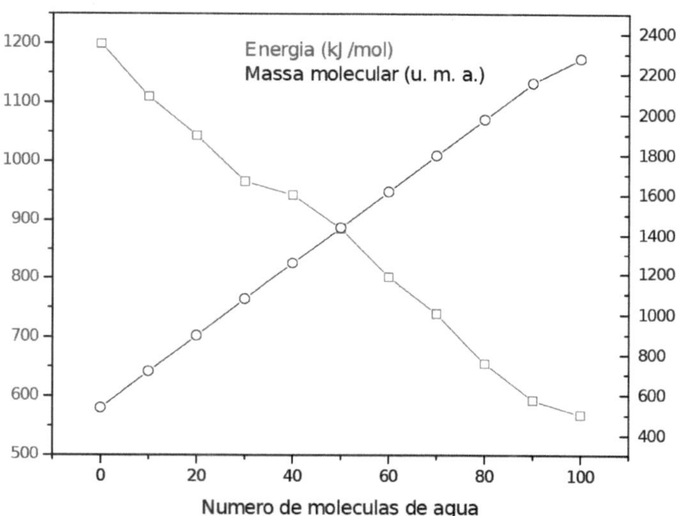

Graph 1 shows the energy and molecular mass as a function of the number of water molecules in its vicinity. This graph shows the different behavior when we add water molecules to the vicinity of the tartrazine molecule. The molecular mass increases while the energy decreases. This is due to the interaction between the water molecules and the tartrazine molecules, where the total energy drops due to the increase in water molecules in its vicinity. See equation 1 below:

$$E = E_t + n E_{H2O} + 1/2(n-1) E_{Dipolo-dipolo} \quad ,$$

where E_t is the energy of the tartrazine molecule calculated in the UFF field; E_{H2O} is the energy of the water molecule also calculated in isolation (UFF field); n is the number of water molecules in the vicinity of the tartrazine; and $1/2(n-1) E_{Dipolo-dipolo}$ is the dipole-dipole energy between the water molecules. This energy obeys the following expression:

$$E_{Dipolo-dipolo} = \frac{-1}{4\pi\varepsilon_0} \frac{p_A p_B}{R^3} (2\cos\theta_A \cos\theta_B - \sin\theta_A \sin\theta_B cos\varphi)$$

Through this expression of the dipole-dipole moment, which is an interaction promoted between the addition of water molecules and the tartrazine molecule, we can explain why the energy decreases in relation to the increase in water molecules in the vicinity of the modeled molecule. With the addition of one water molecule, this interaction is positive, but of a very small order, approximately 10^{-11} kJ/mol. With the addition of 2 or more water molecules, there is already a negative interaction between the water molecules and the modeled molecule, which makes the energy graph decrease.

Graph 2 shows the distance between the C1 and C2 carbons containing the azo group (see Figure 8). We can see that, for the isolated molecule, the distance between carbons C1 and C2 is 3.657 angstroms. However, this value is slightly reduced if we add 10 or 20 surrounding water molecules. This occurs linearly with the increase in water molecules. As soon as there are two water molecules in the vicinity, the energy of the interaction between the water molecule and the modeled molecule (dipole-dipole), which is described by the equation of $E_{Dipolo-dipolo}$, on the previous page, becomes very small and negative. When n = 20 water molecules are added, this distance decreases to 0.11% of its initial value. From n = 20, this distance oscillates around the average value of (3.6545±0.0005) angstroms.

Graph 2: Distance between the $_{C1}$ and C2 carbons containing the azo group.

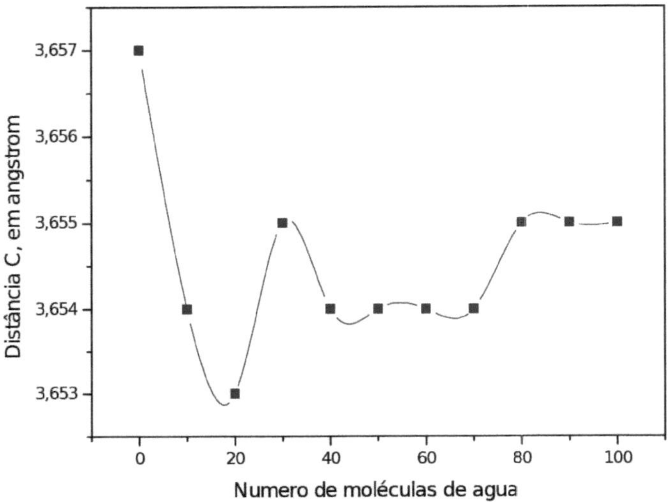

Another important analysis ά that of the angles α and β seen in Figure 8. The graph 3 below shows the behavior of these angles in relation to the increase in water molecules in the vicinity of the tartrazine molecule.

Graph 3: **Alpha and Beta angles in the UFF field.**

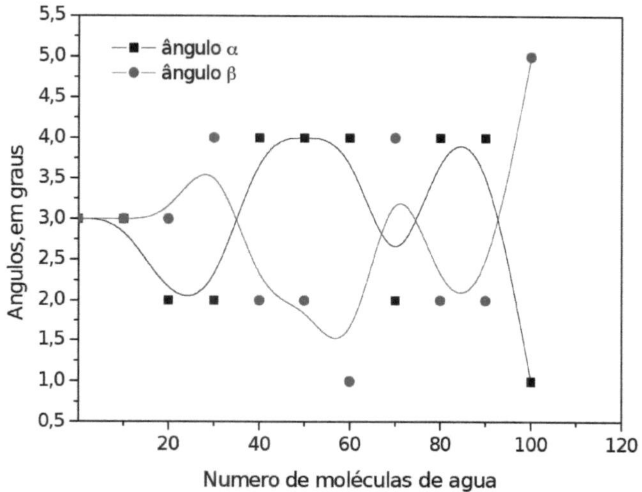

The symmetry seen in Graph 3 was already expected, as the force field adopted disregards some interactions, reducing the adjustment of the atomic radii dependent on the hybridization, an adjustment of the hybridization angles, the parameters concerning the van der Waals interaction, the torsion conformation, and the effective charges of the atoms.

4.3. Ghemical azo group analysis

So far, the results shown have been done using the UFF force field. We will now present the same results using another force field, Ghemical. This force field is characterized by considering molecular interactions, which differs from the first field where some interactions are considered negligible. Table 2 shows the results obtained from modeling tartrazine using the Ghemical force field.

In the UFF field, we saw that molecular mass and energy behave differently, while molecular mass increases, energy decreases. This is seen due to the addition of water molecules in the vicinity of the tartrazine molecule, this behavior is also seen in the Ghemical force field, that is, the two force fields have the same behavior for the analysis of the molecular mass and its energy variation when we are varying its medium, Graph 4 shows this behavior.

Table 2: Tartrazine molecule data (Ghemical force field).

Tartrazine	Energy (Kj/Mol)	Ang link		Molecular mass r (G/Mol)	Distance (-N=N-)	**Distance between** ci **and** c2
		A	β			
Tartrazine (Normal)	112,981	119,4	119,2	534,363	1,346	3,553
With 10 molecules of water	-38,7579	119,0	119,5	714,516	1,345	3,550
With 20 molecules of water	-202,301	119,1	119,5	894,669	1,345	3,549
With 30 molecules of water	-351,873	118,7	120,0	1074,822	1,345	3,538

With 40 molecules of water	-501,688	119,4	119,5	1254,975	1,345	3,553
With 50 molecules of water	-671,135	119,4	119,3	1435,127	1,345	3,545
With 60 molecules of water	-842,303	119,0	120,3	1615,280	1,345	3,548
With 70 water molecules	-1007,35	118,7	120,0	1795,433	1,345	3,548
With 80 molecules of water	-1176,63	118,9	119,1	1975,586	1,345	3,540
With 90 molecules of water	-1320,25	118,9	120,1	2137,723	1,345	3,540
With 100 molecules of water	-1349,43	118,3	119,6	2277,814	1,345	3,540

The interesting thing here is that, even though there are molecular interactions, the behavior of the energy associated with the molecule in the Ghemical field is similar to the behavior of the UFF force field, as we can see by comparing Graph 4.

Graph 4: Energy and molecular mass of the system as a function of the number of water molecules surrounding the tartrazine.

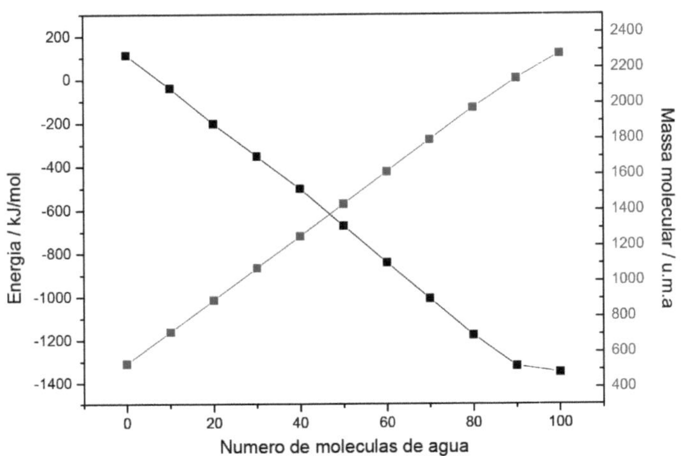

In the graph of the distance C for this field, there is a slight difference with Graph 2 generated by the UFF field. In the UFF field, the molecule is isolated or without interaction, with a distance C of 3.657 angstroms, a result that is quite different from the result obtained in the Ghemical force field, which is 3.553 angstroms. Regarding the behavior of the distance with the increase in the number of water molecules in the medium, for the UFF force field until the addition of 20 molecules the distance C goes decreasing, from n = 20 molecules this distance oscillates around the average value of (3.6545±0.0005) angstroms. In the Ghemical field, the distance C decreases by about 0.4% of its initial value up to n = 30 water molecules, after which it oscillates around 3.544 angstroms.

If we compare the C distances generated by the two force fields without the addition of water molecules, only with the modeled and optimized molecule, we see that the C distance generated by the Ghemical force field is 2.8% smaller, which proves that the molecular interactions affect the conformity of the molecule and consequently in this specific case reducing the azo coupling, which is affecting the chemical instability of the molecule.

The conformity of Graph 2 for the UFF field begins to be maintained for $n \geq 80$ surrounding water molecules, about 0.1 % of the initial value of C. For the Ghemical field, this conformity also begins to be maintained for $n \geq 80$ surrounding water molecules, corresponding to about 0.4 % of its initial value, with a difference of 0.3 % in the distance between the linking carbons that contain the azo group.

Graph 5: Distance between the C_1 and C2 carbons containing the azo group.

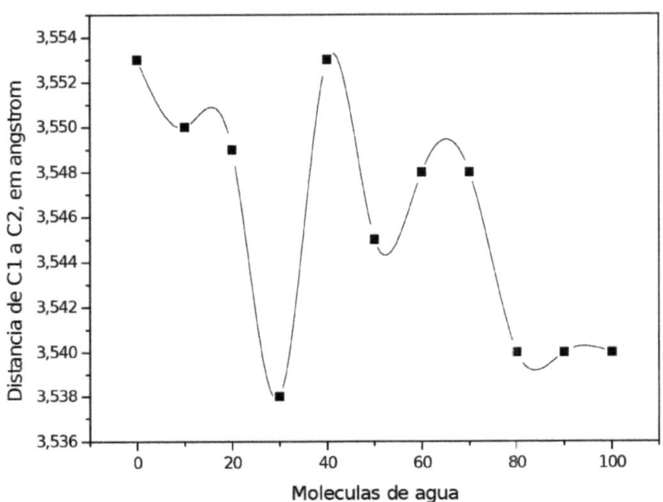

For the angles α and e, carbon-bonding angles seen in Figure 8, the graph of the UFF field showed an expected symmetry, as this field disregards some interactions between the molecules. In Graph 5, which considers the angles α and β generated by the Ghemical field, the behavior is symmetrical, but the difference is that the angle α has increasing symmetry in relation to the angle β. With the increase in water molecules, the α angle is increasing in relation to the β angle, thus affecting the structure of the molecule and modifying the natural behavior of the azo group in relation to electronic delocalization, deregulating the absorption of light in the visible region, which will consequently affect the color of this dye.

Graph 6: Alpha and Beta angles

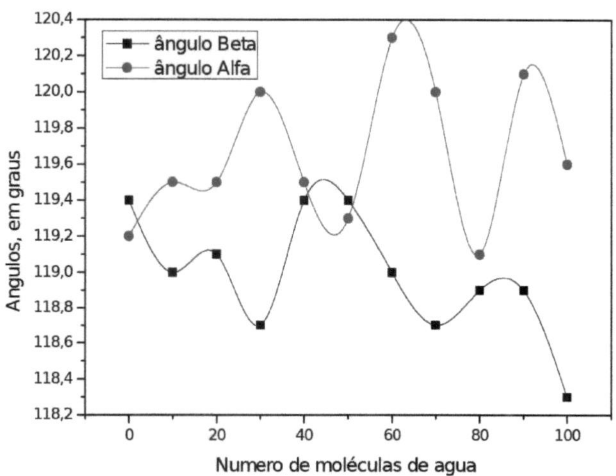

One of the major concerns regarding the use of azo dyes in food is maintaining their chemical stability. By comparing the analysis of the azo group in the UFF and Ghemical force fields seen in the results cited above with the comparison between graphs 1, 2, 3, 4, 5 and 6, the behavior of tartrazine in an aqueous environment can be evidenced and simulated. With this comparison we can analyze that the azo group, the group responsible for coupling the chromophore group, behaves in a range of n = 0 and n = 100 surrounding water molecules. Looking at the graphs with the parameters analyzed in Figure 4, we can see some differences between the force fields, which reinforces the idea that due to the complex diversification of food matrices, the chemical environment in which the dye will be exposed is one of the main factors impacting on the maintenance of the molecular structure of the chromophore and, consequently, on the regularity of the color in the food.

Chapter 5: Conclusion

The comparison between the UFF and Ghemical force fields shows the impact that a chemical environment can have on the molecular structure of the molecule as a whole. In the case of this work, only the behavior of the azo group of the Tartrazine molecule was analyzed. As it is a highly soluble molecule in an aqueous environment, we checked the behaviour of the azo group in relation to the force fields. As there are no interactions between the molecules, the UFF field does not destabilize the molecule structurally, which is quite different from what happens with the Ghemical force field, as it modifies the structure of the molecule, it reduces the bonds of its atoms, consequently promoting a greater dipole moment between its atomic bonds, promoting a symmetrical angle, which was also seen in the UFF force field, but the difference is that the angle α increases in relation to the angle β, affecting the structure of the chromophore group, and as already mentioned, the regularity of the color of this dye.

Chapter 6 : References

1 At http://www.chemheritage.org/discover/online-resources/chemistry-in-history/themes/molecular-synthesis-structure-and-bonding/perkin.aspx (accessed December 19, 2014).

2 Old.iupac.org/publications/cd/medicinal_chemistry/Practica-in-3.pdf (accessed June 14, 2014)

3 PINHEIRO,H.M.;TOURAUD,E.;THOMAS,O.Aromatic amines from azo dye reduction:status review with emphasis on direct UV spectrophotometric detection in textile industry wastewaters. Dyes And Pigments,v. 61,n. 2,p. 121-139, 2004.

4 TOPLISS, JOHN G., Molecular Modeling in Drug Design, Journal of Medicinal.

5 At http://openbabel.org/wiki/OBForceFieldGhemical. (Accessed June 10, 2014).

6 http://quimicaestrutural.blogspot.com.br/2010/04/aprendendo-um-pouco-mais-sobre- quimica.html (accessed June 1, 2014).

7 NUSSENZVEIG, H. M., Curso de Fisica Bàsica, vol. 3, Eletromagnetismo, 1ª edição (6ªimpressão, 2007), Blucher, Sao Paulo, Brazil, 1997.

8 GRIFFITHS, D. J, Electrodynamics, 3rd edition, Pearson, Sao Paulo, Brazil, 2011.

9 http://efisica.if.usp.br/moderna/mq/sistemas_compostos/ (accessed March 30,

2014).

10 http://avogadro.openmolecules.net/wiki/Main_Page. (Accessed February 25, 2014).

11 http://mathfaculty.fullerton.edu/mathews/n2003/GradientSearchMod.html (accessed February 15, 2014).

12 http://mathworld.wolfram.com/ConjugateGradientMethod.html (accessed April 15, 2014).

13 http://avogadro.openmolecules.net/wiki/Main_Page (accessed March 30, 2014)

14 NASCIMENTO, VALTER A.; MELNIKOV, PETR; CONSOLO, LOURDES Z. Z., Computerized Modeling of Adenosine Triphosphate, Adenosine Triarsenate and Adenosine Trivanadate, *Molecules*, 17, 2012.

Printed by Books on Demand GmbH, Norderstedt / Germany